essentials

Springer essentials

Springer essentials provide up-to-date knowledge in a concentrated form. They aim to deliver the essence of what counts as "state-of-the-art" in the current academic discussion or in practice. With their quick, uncomplicated and comprehensible information, *essentials* provide:

- an introduction to a current issue within your field of expertise
- an introduction to a new topic of interest
- an insight, in order to be able to join in the discussion on a particular topic

Available in electronic and printed format, the books present expert knowledge from Springer specialist authors in a compact form. They are particularly suitable for use as eBooks on tablet PCs, eBook readers and smartphones. *Springer essentials* form modules of knowledge from the areas economics, social sciences and humanities, technology and natural sciences, as well as from medicine, psychology and health professions, written by renowned Springer-authors across many disciplines.

Martin Wiedemann

System Lightweight Design for Aviation

Springer

Martin Wiedemann
Braunschweig, Germany

ISSN 2197-6708   ISSN 2197-6716 (electronic)
essentials
ISSN 2731-3107   ISSN 2731-3115 (electronic)
Springer essentials
ISBN 978-3-031-44164-6   ISBN 978-3-031-44165-3   (eBook)
https://doi.org/10.1007/978-3-031-44165-3

This Springer imprint is published by the registered company Springer Nature Switzerland AG
The registered company address is: Gewerbestrasse 11, 6330 Cham, Switzerland

The paper of this product is recyclable.

What you can find in this *essential*

- A definition of lightweight system design as an extension of classic lightweight design
- An overview of the possible energy sources of the future in aviation
- An assessment of the importance of weight and drag reductions for commercial aircraft energy consumption
- An insight into the potential of design concepts using carbon-fibre reinforced polymers
- An overview of research results in lightweight system design in the fields of materials, methods, design concepts, and manufacturing technologies to achieve the climate goals of minimum-emission aviation

Foreword

To achieve the climate goals of the Green Deal, commercial aircraft are expected to fly on new energy sources whose availability and manufacturing costs make it necessary to reduce energy consumption. In addition, even when hydrogen or synthetic fuel is burned, a certain amount of residual emissions will always be produced. Lightweight system design makes decisive contributions to an ecological and economical aviation of the future, both in terms of weight savings and drag reduction.

Fibre composite materials are increasingly being used in aircraft structures. The performance capabilities of this class of materials are great and have not yet been exhausted. Further weight savings are made possible by integrating functions into the fibre composite structure. Fibre composites are also suitable for reducing aerodynamic frictional drag. How will weight savings and drag reduction affect the energy consumption of future aircraft? What distinguishes lightweight system design from conventional lightweight design? Which methods, design concepts, manufacturing technologies and which functional integration options are available for energy-efficient aircraft of the future?

This *essential* is aimed at interested engineers and experts in the aviation industry who are looking for solutions that can be implemented quickly to increase the efficiency of future commercial aircraft. The overview is intended to illustrate that lightweight system design enables savings in weight of more than 10%, cost reductions in manufacturing and significant contributions to aerodynamic drag reduction.

After an introductory overview of expected costs of future energy carriers for aviation, a specification of the potentials of weight and drag reduction for energy consumption, emissions and costs and an introduction to the systematics, exemplary results from research are summarised thematically sorted with short introductory explanations. The short descriptions with references allow interested readers a quick and easy insight into further details.

The cited research results represent only a small portion of the large pool of possibilities of lightweight system design to contribute to an emission-minimised aviation of the future. In addition, the *essential* aims to stimulate further research in the direction outlined.

Martin Wiedemann

Contents

Motivation

Alternatives to kerosene as an energy source are currently being discussed for minimum-emission—ideally zero-emission—aviation. However, their availability is likely to be limited, they will be expensive, and, according to current knowledge, they will not be completely emission-free. Energy-saving technologies for future commercial aircraft will therefore play a major role in achieving the goal.

This is where lightweight system design comes in, as it enables further weight savings and significant contributions to the reduction of aerodynamic drag beyond conventional lightweight design through function integration.

Carbon-fibre reinforced polymers (CFRP) are particularly suitable for lightweight system design due to their high structural performance and their possibilities for function integration.

1.1   Lightweight System Design for Low-Emission Aviation

▶ Definition
Lightweight system design is an extension of classic lightweight design. In lightweight system design, the aim is to integrate as many passive and active functional elements as possible into the load-bearing airframe.

In aircraft design, such functional elements include aerodynamic fairings, cabin fittings, electrical lines, antennas and energy storage systems.

In addition, technologies for aerodynamic flow control can be integrated into lightweight system design to help reduce aerodynamic drag.

Lightweight system design is understood as lightweight design in interaction with other systems of the aircraft.

© Deutsches Zentrum für Luft- und Raumfahrt e. V. (DLR), Linder Höhe, 51147 Köln 2024
M. Wiedemann, *System Lightweight Design for Aviation*, essentials,
https://doi.org/10.1007/978-3-031-44165-3_1

CO_2 emissions in aviation are mainly caused by three classes of aircraft: regional aircraft 7%, short-/medium-range (SMR) aircraft 51%, and long-range (LR) aircraft 42%. [38]. Weight savings and drag reduction in SMR and LR aircraft are therefore of particular interest.

Possible aviation energy sources include green-energy-based synthetic fuel, also known as e-fuel, liquid hydrogen (LH2), methane, ammonia, liquid organic hydrogen carriers (LOHC), and batteries. An overview is given in Table 1.1.

Costs of kerosene, e-fuel, LH2, methane [4], costs ammonia [5], primary energy use [1, 2], battery vol. energy density [32], battery grav. energy density [16, 64], battery costs [64].

e-fuel is attractive because it would enable even today's aircraft to fly CO_2-neutral. However, at significantly higher costs and with high primary energy input.

LH2 is two thirds lighter than kerosene in terms of gravimetric energy density, requires four times the storage volume of kerosene, is cheaper and requires less primary energy input than e-fuel, but is still more expensive than kerosene.

Methane, ammonia, dibenzyltoluene (LOHC), and batteries tend to be out of the question for medium- to long-range aviation because of their lower volumetric and gravimetric energy density.

Table 1.1 Possible energy sources for aviation

	Vol. e-density	Grav. e-density	Costs 2020	Costs 2030	Costs 2050	Primary energy use
	[kWh/l]	[kWh/kg]	[ct/kWh]			In/Out
Kerosene (incl. EUA)	9.7	11.9	5.7	7.4	10	
e-Fuel	9.7	11.9	40	35	28	2–5
LH2	2.36	33.3	22	19	16	1.35–2
Methane (liquid)	4.42	10.8	29	29	23	1.72
Ammonia (liquid)	4.25	6.25	14–22	17–28	15–23	1.6–2
LOHC	2	2		14–25	12–20	1.35–2
Lithium battery (target)	0.35	0.5	7–11	5–8		1

A particular challenge when using LH2 is the resulting additional weight from the tank and piping system.

For example, for the Ariane 6 with a single-use tank, Air Liquide Energies has developed a metallic LH2 tank that holds 28 tons and weighs 5.5 tons [9]. The effective storage density is thus 28 kWh/kg.

In a DLR study [104], a total system weight of 2432 kg is assumed for an LH2 filling weight of 781 kg when a CFRP tank is positioned in the rear of the aircraft. The resulting effective storage density of 10.7 kWh/kg is on a par with kerosene, but the additional volume remains.

The required quantities of (green) e-fuel or hydrogen will be a particular challenge, in addition to the cost price and remaining residual emissions, as aviation competes with other consumers [23].

Whatever energy source is chosen or will be available in the future: it remains crucial to reduce energy consumption. To achieve this goal, there are two main influencing factors: weight reduction and drag reduction.

1.2 Potentials of Lightweight System Design

With regard to possible weight savings, a distinction is made between primary and secondary structures (Fig. 1.1). The former describes the load-bearing airframe structure and must therefore meet particularly high safety and certification requirements. Improved methods of classic lightweight design are predominantly used for primary structures. Weight savings in secondary structures and cabin elements as well as passive and active systems to support aerodynamic laminar flow are in the focus of lightweight system design.

The primary structure of a typical SMR aircraft (wings, fuselage and tail units) weighs about 15 tons in today's metallic design. It can be shown that improved knowledge of modern lightweight materials, their structural properties and new design concepts enable a weight reduction of around 20%. From the secondary structures, the systems and the cabin (in total about 13 metric tons), the principle of function integration of lightweight system design can achieve further savings of about 10%. The unladen weight of an SMR aircraft of 44 metric tons (including landing gear, engines and operating equipment) can therefore be reduced by at least 4.3 metric tons with the current state of knowledge. Further weight-saving possibilities lie in active flight control systems for load reduction [97].

For a SMR aircraft, reducing the take-off weight by one ton for a cruise length of 2000 nautical miles, according to Brequet [3], means fuel savings of 171 L of kerosene or e-fuel. Aircraft in this class are expected to fly 60,000 flight cycles

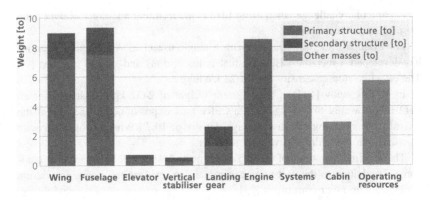

Fig. 1.1 Typical weight distribution for a short- and medium-range (SMR) aircraft

[11]. The reduction in take-off weight thus saves 10.3 million litres of fuel over the course of an aircraft's lifetime. This does not take into account additional savings from the reduced fuel weight. Assuming 100% availability of e-fuel in 2050 and an optimistic price of 0.2 €/kWh [4], one ton of take-off weight will save almost €20 million during the aircraft's lifetime. If the use of fossil fuels is assumed to continue, this results in a saving of 25,800 t CO_2 for the example considered, and since the price of kerosene will also increase by 2050 due to higher production costs and emission certificates, at least €10 million will also be saved in this case.

A major contribution to minimising the energy consumption of future aircraft is made by lightweight system design in conjunction with aerodynamics in reducing frictional drag. By avoiding waviness and edges the surfaces, natural laminar flow (NLF) is enabled at maximum run length over the airfoil, and active laminar flow control (LFC) is supported with the aid of boundary layer vortex extraction and variable control of flow separation at the lift surfaces and fuselage.

According to Brequet, a 1% increase in glide ratio will result in fuel savings of 9 million litres of fuel over the lifetime of an SMR aircraft based on the above assumptions, and operating cost savings of €18 million for e-fuel.

If kerosene is used after all due to the lack of availability of e-fuel, improving the glide ratio by 1% would save around 22,500 t CO_2 and reduce operating costs by €9 million.

It goes without saying that the costs of manufacturing lightweight airframe components must also be kept within limits. Here, too, developments in lightweight system design in the field of manufacturing technologies and quality assurance offer significant potential for savings; often not against but with simultaneous weight savings.

1.3   The CFRP Lightweight Material

One possibility for lightweight system design is the substitution of materials towards higher structural performance. The use of CFRP in particular offers great potential in terms of

- weight, due to the good specific strength and stiffness,
- function integration, due to the inherent merging of multiple material components,
- laminar holding, due to possibilities of step- and gap-free integral design.

Fibre composites (FC) made of CFRP are characterised by the highest (weight-)specific strengths and stiffnesses, see Fig. 1.2. This fact, combined with the corrosion resistance and very high fatigue strength, has led to an increase in the share of CFRP in the primary structure of civil aircraft of the last generations to about 50%.

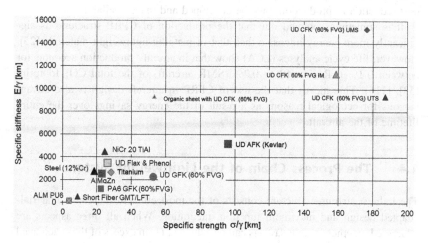

Fig. 1.2 Comparison of the specific strength (tensile length) and specific stiffness of various lightweight materials

If strength and stiffness in one load direction alone were the determining factors for an aircraft structure, CFRP structures would weigh only 20% of a comparable light-metal structure for the same load-bearing capacity. However, this very high lightweight potential cannot nearly be exploited for many reasons. One reason is that the fibers must be oriented in multiple load-bearing directions; this results in at least orthotropic layup of plies. Another reason is the lack of plasticity, which makes damage barely visible. When subjected to impact loads, CFRP structures may show delaminations in the resin area that cannot be visually detected, which reduce the load-bearing capacity, especially under compressive and shear loads. Therefore, structural properties are often set far below the theoretically possible material properties (Sect. 2.2). Another major difficulty lies in the non-fibre-composite suitable design of today's aircraft structures. For example, current airworthiness requirements restrict the approval of structural bonding in the primary structure (Sect. 2.3) and structural bonding (Sect. 2.4).

Removing these and other limitations in the use of CFRP is of great importance for future lightweight structures. Results from research show that significant progress is possible here.

The use of CFRP in wing structures is now the rule. The use of CFRP in fuselage structures, on the other hand, is not yet established, as it has been shown that the weight advantage is not as high as initially predicted for the reasons mentioned above, among others, and the manufacturing costs are significantly higher than those of comparable metallic structures. Therefore, further possibilities for cost reduction in production must be developed and made available.

It is sometimes pointed out that the production of CFRP structures is significantly more energy-intensive than that of metallic lightweight materials [25]. However, life cycle analyses (LCA) show that in aircraft, production accounts for between 0.1% (LR aircraft) to 0.2% (SMR aircraft) of the total CO_2 footprint [53], [113]. This means that the use of CFRP in aircraft design is also advantageous for ecological reasons as a result of the energy savings over the entire lifetime of the aircraft.

1.4   The Process Chain of the Lightweight System

The lightest airframe structure consists of the most efficient materials, material-adapted design and maximum function integration. When all three aspects are addressed together, weight savings can be achieved far in excess of those achieved today.

Fibre composites belong to the class of generative materials, whose mechanical properties only arise in the manufacturing process of the components from different semi-finished products. A characteristic feature is the interaction between materials, simulation tools, design concepts, manufacturing methods, integration of functions and implementation on an industrial scale. New calculation methods enable new design concepts and new manufacturing technologies enable the use of new materials. Individual findings can be combined in very different ways to create lighter, lower-drag, lower-cost, and easier-to-maintain aircraft structures. In lightweight system design, even initially inconspicuous findings can have a major impact on the way to the airframe.

A look at the process chain and exemplary research results in the subareas is helpful in understanding the basis of the potential estimates for weight savings and drag reduction in lightweight system design. This results in a wide variety of options for the aviation industry along the entire value chain of aircraft design. In the process chain, Fig. 1.3, adaptronics exploits the potentials of function integration from a systemic perspective.

Even if the explanations in the following chapters strive to do so, not all of the research results described here can be directly quantified in terms of their effect on weight, manufacturing costs or drag reduction because, for example, they have different effects in interaction with different aircraft concepts and design methods, but also semi-finished products or manufacturing technologies.

The examples are grouped into three main chapters:

- Contributions from classic lightweight design
- Lightweight system design with integration of passive functions
- Lightweight system design with integration of active functions

Most of the results reported in the next chapters point to potentials that have not yet been realised. The selected examples almost all originate from research work carried out over the last 15 years at the DLR Institute of Lightweight Systems

Fig. 1.3 Process chain of lightweight system design

together with partners. A large number of further research results on lightweight system design can be found, for instance, in [6].

The technological topics addressed in the examples are **highlighted in underline** in the following chapters to make them easier to find.

Classic Lightweight Design

2

Along the process chain, Fig. 1.3, new materials, better material properties, improved and more accurate design methods, new design concepts, new joining technologies, more efficient production technologies, and new automated quality assurance processes make diverse contributions to optimised and more cost-effective lightweight design. The basis is carbon-fibre reinforced polymers (CFRP), Sect. 1.3, whose lightweight design potential can be exploited much more extensively with the research results cited below as examples.

2.1 New Material Hybrids and Semi-Finished Products

The development of new lightweight materials is characterised by strong dynamics. There are currently a large number of knowledge platforms and research programmes in materials development in Germany. The potentials are manifold, as the following examples show, but only sufficiently high advantages justify the extensive qualification efforts required in aircraft design before introduction in primary structures.

Material hybrids are a combination of different material classes. The material classes referred to here are metals, fibre composites and different matrix materials. Metal-fiber hybrids are more in the performance range of metals (compare Fig. 1.2), but by combining their properties, they can be advantageous in many applications where strength and stiffness are not dimensioning factors.

Fibre-metal laminates (FML, Fig. 2.1) utilise the specific properties of metals (isotropy, ductility, electrical conductivity) in combination with those of fibres (high strength and stiffness, no fatigue, no corrosion). The best-known fibre-metal

© Deutsches Zentrum für Luft- und Raumfahrt e. V. (DLR), Linder Höhe, 51147 Köln 2024
M. Wiedemann, *System Lightweight Design for Aviation*, essentials,
https://doi.org/10.1007/978-3-031-44165-3_2

Fig. 2.1 Structure of a fiber-metal laminate

laminate is GLARE (fibreglass plies with aluminium foils) used in the A380, developed to increase the fatigue and residual strength of aluminium. In contrast to GLARE, however, the use of aluminum should be consistently avoided when using carbon fibers. Nevertheless, FMLs have many advantages for lightweight design.

For CFRP tensile specimens with inserted titanium foils, up to 30% higher bearing strength can be demonstrated compared to a pure CFRP laminate [24], and with inserted steel foils a 15% higher strength against compression load after impact (CAI) [106]. When metallic foils are inserted in the outer layers of a laminate, a 50% increase in mass-specific energy absorption can be demonstrated [21].

Further applications of fibre-metal hybrids are discussed in Sect. 3.2.

Hybridisation of FC is also possible with elastomers (ethylene propylene diene monomer: EPDM) and is useful for various applications. An FML with absorber layers of EPDM exhibits an up to 35% smaller delamination area at impact compared to pure CFRP laminate [29, 30]. Furthermore, elastomer-CFRP hybrids are very effective in the area of shape variability (morphing), Sect. 4.2.

Matrix **modifications** improve the properties of epoxy resins, for example. To minimise matrix-based failure mechanisms in FC such as crack growth and delamination, nanoparticles are added to the resin. Basically, the smaller the particles, the larger the specific area and the better the fracture toughness and energy release rate. Other mechanical properties can also be positively influenced with nanoparticles.

For example, boehmite (aluminum hydroxide) can increase the fracture toughness of an epoxy resin by 39% and the energy release rate by 66%, as well as the stiffness of the resin by 17% [54]. Mineral nanoparticles with 30% filler content increase the thermal conductivity by up to 40% and reduce the fire spread rate to one third [67]. Taurine-modified boehmites with 15% filler content improve not only the E-modulus, fracture toughness and energy release rate but also the fatigue behaviour of the resin [75].

Thin ply laminates can be produced by special spreading technology with basis weights of 20 g/m^2 to 100 g/m^2 compared to standard laminates with 160 g/m^2 to 600 g/m^2 (Fig. 2.2). Coupon specimens exhibit better mechanical properties, better fatigue behaviour and smaller delamination areas [13] as well as increased resistance to cracking [98]. Omega stringers made from thin-film laminates also exhibit significantly smaller delaminations than those made from standard laminates [22]. A good overview of the potentials of thin-film laminates and the challenges of their processing is given in [88].

High-temperature **CFRP materials** offer the advantage of load-bearing capacity even at elevated temperatures, which can save heavier protective materials.

Studies conducted as part of an EU project promise up to 15% weight and up to 17% cost reduction potential and give a selection of possible matrix systems

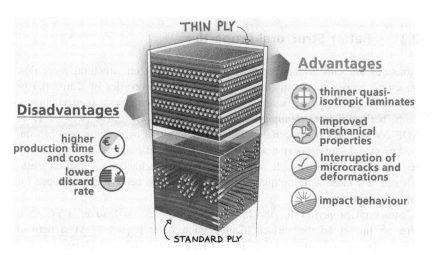

Fig. 2.2 Thin film laminates—advantages and disadvantages

[121]. Calculations show that a stiffened CFRP shell with conventional epoxy matrix still carries 53% of the load under axial pressure at 210 °C compared to the shell at room temperature [72].

Long-fibre-reinforced polymers have lower lightweight potential compared to continuous-fibre-reinforced polymers, but can be used for secondary structures or cabin applications (cf. Fig. 1.1), and can be well adjusted in terms of their stiffness by aligning the long fibres (up to 25 mm in length). The longer the fibres, the better the load transfer between fibres through the matrix material. Using long-fibre GFRP as an example, it can be shown that with the same orientation and fibre volume content (FVC), 50% of the stiffness can also be achieved [65] compared to a composite with continuous fibres.

Recycled materials and natural fibres are becoming increasingly important in secondary structures for aircraft design. Currently, recycled carbon fibres (rCF) or offcuts from manufacturing are used in various applications. These are long fibres, which are mostly processed in the form of non-wovens with a rather low FVC of about 30%. They also allow the addition of natural fibres. Composites of rCF non-wovens and epoxy matrix with 30% FVC have about 170% of the specific strength and 70% of the specific stiffness of lightweight aluminium (AlMgZn) [15]. Pure sisal fibres show up to 22 GPa stiffness and a composite with 30% FVC of 75% flax fibers and 25% rCF shows a stiffness of 12 GPa. These and other results were recently obtained in the EU-China project ECO-COMPASS [14].

2.2   Better Structural Properties

Structures are dimensioned against failure depending on structural properties. Structural properties are often not the mechanical properties of a material or a structure (strength, stiffness, etc.) measured on specimens or (better) components, but result from multiplying them by knock-down factors (KDF < 1). A KDF compensates for uncertainties in the description of a failure mechanism, load or manufacturing deviations. Often, a mechanical property is multiplied by several KDFs. More accurate and reliable failure mechanism calculation methods and new manufacturing quality assurance methods need to be developed for future weight savings.

Structural properties for the longitudinal compression load of a FC shell after an impact on the surface (Compression After Impact—**CAI structural properties**) often destroy lightweight construction potentials of high-performance CFRP semi-finished products. This is due to possible delaminations as a result of the impact, which cannot be detected visually due to lack of plasticity. Today, the results of simple CAI coupon tests are directly transferred to a real structure. However, shells react differently to impact: if a thin shell is hit in the middle between longitudinal and transverse stiffeners, part of the energy is compensated elastically and does not turn into damage (delamination); if, on the other hand, the impact hits a rear-supported shell, the damage is much greater. Today's CAI coupon tests do not take this variance into account. The elastic energy component of an impact can be determined by new methods and differentiated with respect to the CAI characteristics on component level, thus allowing effective weight savings [20]. On the basis of simple calculation approaches, different designs against impact damage can already be performed at the preliminary design level. An overview of the known approaches for the consideration of impact damage on FC shells and recommendations for simple application are given in [19].

The **fatigue strength of** FC is significantly better than that of metals, but the proof for a whole aircraft lifetime is not easy to provide. In order to investigate the fatigue strength of a FC, resonant test methods are required that realise a number of load cycles $>10^7$ at a high frequency in a reasonable time. The test specimens must be suitably selected to avoid unrealistic boundary effects and to ensure cooling of the specimens to a constant temperature. For a GFRP-epoxy composite, fatigue strength up to $2 \cdot 10^7$ load cycles was demonstrated in a resonant test procedure for a strain level of 1700 micro strain [74].

The **permissible strains of FC** are currently restricted by KDFs due to the poor detectability of damage. SHM systems (Structural Health Monitoring) make invisible damage in the structure detectable, Fig. 2.3. Piezoceramics can emit and receive guided ultrasonic waves (confined by the surfaces of the structure) over a wide area. These interact with changes in stiffness or delaminations, for example. Damage can be detected and localised by comparing the TARGET-ACTUAL sensor signals. Thanks to such systems, the KDFs for allowable strains of FC can be increased and the structure can be built thinner. Weight savings of at least 5% could be demonstrated using the example of a vertical stabiliser, taking into account the SHM system weight [27].

2.3 New Design Concepts

In order to build lighter with CFRP, fibre-composite-compatible design methods are required. These include a load-oriented fibre placement, the use of the anisotropy of a laminate structure and the application of bonding technologies also for primary structures. The advantages of integral design and relatively free

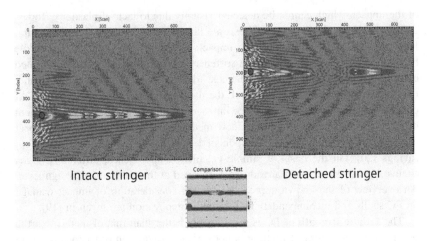

Intact stringer   Comparison: US-Test   Detached stringer

Fig. 2.3   Effect of guided ultrasonic waves on damage detection

shaping should be considered early in the design process. Practically, there are some challenges in the implementation and use of such design concepts. Here, with a view to the further development of calculation tools and manufacturing technologies (digitisation), it is worthwhile re-evaluating already known concepts.

A **fibre-composite suitable design** saves weight and manufacturing costs. An overview of requirements and different FC fuselage concepts was developed in a LuFo project (German aeronautical research programme). For example, a sandwich (SW) skin-shell concept with integrated longitudinal stringers and continuous frames can save up to 30% compared to the current A320 fuselage, Fig. 2.4. If – in a different concept – the cargo compartment is located outside the pressurized cabin[55]there is a potential of 25% weight savings compared to an A320 [63].

The **surrounding structure of a passenger door** in an aircraft fuselage is heavy because of the stiffening of the large cutout against shear deformations and the support of locally concentrated loads from the door fittings, and also expensive because of the complex structure. In the 7th Framework Programme of the EU, the project MAAXIMUS. [47] new possibilities of a fibre-composite, high-integrity door-frame structure (Door Surround Structure: DSS) were investigated. Considerable weight savings can be achieved in terms of shear stiffening and skin thickening in the corners of the frame [118]. An integral design method has also

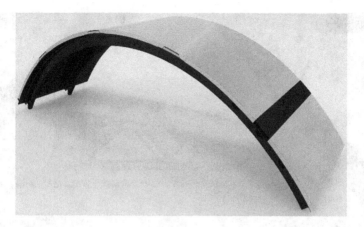

Fig. 2.4 Integral fuselage segment consisting of longitudinal stiffeners (longerons) and sandwich panels

been developed, which allows to effectively reduce the cost in the assembly of a door frame [62].

Weight, design complexity and costs can be saved if airframe shells are designed according to the double-double (DD) laminate principle. In particular, doubler runouts can be significantly reduced, which saves weight [57].

Ultra-lightweight design for special applications, for example high-flying unmanned communication platforms, reveals the limits of what is possible. For a high-flying platform with a maximum take-off weight (MTOW) of 135 kg and a wingspan of 30 m, a construction of a tube spar in winding technology with airfoil-forming sandwich ribs and covering with a total wing area weight of 0.9 kg/m^2 is possible, Fig. 2.5, [116].

2.4 New Joining Technologies

The joining of fibre composite components deserves special attention in lightweight design. Process-induced deformations pose a challenge: in the case of structural bonding, the joining partners must fit together precisely over the entire contact surface. If riveted, this results in additional weight and additional costs: The CFRP shells have to be designed with additional layers e.g. weight in

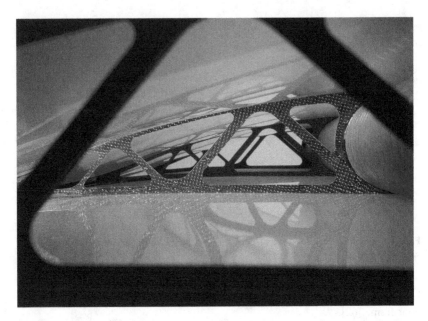

Fig. 2.5 Ultra-light wing structure of a high-flying platform

the area of holes for fasteners due to reduced bearing strength and the required corrosion resistant titanium rivets increase the costs.

In the case of profile-like components such as frames with angular cross sections, **process-induced deformation** (PID) occurs, Fig. 2.6, and, in the case of planar laminates, warping. A calculation of these deformations resulting from the thermal shrinkage of the matrix allows for compensation in the mould. Remaining residual stresses (**process-induced stresses –** PIS) can be minimised by analysing the curing behaviour and suitable temperature control.

Efficient simulation methods for precise target contour setting, based on tests of representative small samples (coupons), are available [58] and demonstrated on the example of a complex CFRP structure Fig. 2.6) for the compensation of the validated PID [56]. The effect of process parameter scatter on PID and PIS can also be determined [71].

Structural bonding is still only possible to a limited extent for primary structures in aircraft design because defects in the bond cannot be detected using non-destructive testing methods (NDT). The Acceptable Means of Compliance (AMC) regulations of the aviation authority EASA (AMC 20–29), [31] require

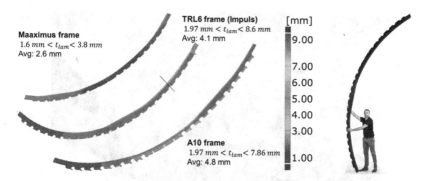

Fig. 2.6 Frame profiles of an aircraft fuselage without and with PID compensation

certification of an adhesive bond to be substantiated by tests for all load types and a clear and individual test of each bond against weakening (weak bonds). A process-safe and certifiable structural bond, based on a special surface activation (Fused Bonding) (Fig. 2.7) and a testing technology for the bond (Bondline Control Technology: BCT) [43], has now been developed for the assembly of primary structures [42] and vividly described in a YouTube video [44]. This process also allows for the bonding of metallic structures.

Repair bonding is subject to increased requirements as it often has to be carried out under conditions with limited control. Several processes are now available to enable adhesive repairs.

Process-reliable adhesive repair is possible with BCT using metallic meshes to test adhesive pretreatment [43]. A hybrid adhesive bond of epoxy with thermoplastic phase stops the propagation of delamination at levels up to 5000 micro strains [73].

2.5 New Manufacturing Technologies

This chapter describes new manufacturing technologies that enable the accurate and cost-effective realisation of new lightweight structures.

Continuous fibres are deposited in flat components by automated tape laying (ATL) or automated fibre placement (AFP) processes with pre-impregnated fibre layers (prepregs) or in dry deposition with subsequent resin infusion (LCM). AFP with narrow fibre tapes allows for the deposition of radii. For complex linear components in profile form with variable cross-sections, dry deposition of textile

Fig. 2.7 Fused bonding – detailed view of the peeling process of the activation foil

semi-finished products with injection in closed moulds (resin transfer moulding: RTM) is usually used. Pultrusion and winding processes are also used.

For dry deposited fibres, impregnation is achieved by different variants of infusion or injection [49].

Different compacting methods (pure ambient pressure, mould, autoclave) are used to generate a defined fibre volume content (FVC). Up to approx. 65% FVC, complete impregnation and enclosure of the fibre filaments with resin is possible.

Manufacturing costs are influenced by the efficiency of fibre deposition, by handling, or better avoiding of manufacturing deviations, and by minimising the consumption of auxiliary materials and energy used.

A good overview of the current status of efficient and cost-saving CFRP manufacturing technologies is given in the project report EFFEKT (LuFo V-2) [60].

Near end-shape **RTM technologies** (Fig. 2.8) offer high automation potential and many advantages in the production of complex components. They can be taken directly from the fixed mould cavity without time-consuming and costly postprocessing such as edge sealing. Isothermal process control allows for minimisation of the required energy input and the use of cheaper mould metals, whose thermal expansion coefficient does not have to be minimised, as is the case with expensive INVAR steels.

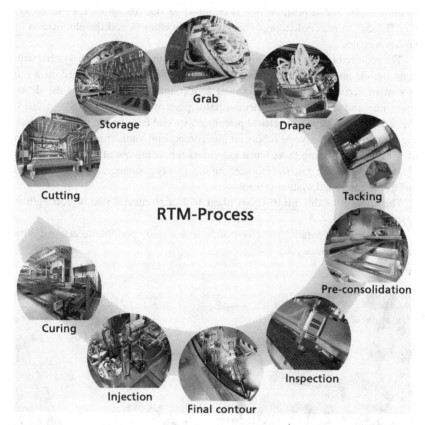

Fig. 2.8 RTM technologies – process chain with individual processes

An overview of new sub-technologies for the efficient application of RTM and advantages in interaction with Industry 4.0 is presented in [117]. Flexible manufacturing concepts and high levels of automation in the use of RTM technology even for low volumes have also been tested [110].

A process time reduction of 60% through isothermal process control and other advantages have been demonstrated using the production of fuselage frame segments [96].

To ensure fibre placement within aerospace tolerances (0.1 mm maximum gap between deposited fibre tapes), mostly rigid gantry systems are used today in which an AFP or an ATL depositing head places fibre material on horizontal forming tools. The deposition rate is currently around 10 kg/h (AFP) to 20 kg/h (ATL) due to material storage changes, quality controls and the elimination of laydown errors.

The required tolerances can also be achieved with robots. Such a system with eight mobile and coordinated robot units, which places fibres arranged on a rail system in vertically positioned forming tools, significantly increases the deposition rate and saves space. A demonstrator plant was set up in Stade in 2013, Fig. 2.9, [7]. In addition to vertical placement, special features are realized including parallel work of several robots on one component, uninterrupted continuation of work by transferring tasks when individual robots are out of phase (for example, for changing the material storage), online quality assurance, and collision-free and event-controlled system control.

The concept of this **multi-robot plant** with a placement rate of 150 kg/h is described in [66].

With a second, parallel placement unit, it is already possible to achieve 30% higher placement rates [26].

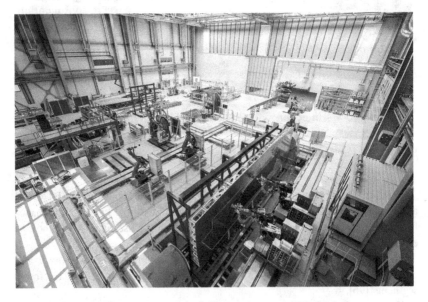

Fig. 2.9 Plant for multi-robotic continuous fibre placement at DLR Stade

Path planning and collision-free distribution of the work orders to the individual robots are *essential* [100]. The multi-robotic fibre placement system in action can be seen here: [122].

The fibre-metal hybrid GLARE has already been mentioned (Sect. 2.1). **Automation of GLARE production** has been developed to reduce manufacturing costs and times. Layer by layer, glass-fibre UD layers are positioned by robots in the correct angular position, thin aluminium foils of up to 1.3 m width and 2 m length as well as adhesive films are deposited without wrinkles in the correct position, and consolidated together with stitched stringers in the autoclave. The production rate could be increased by a factor of 5 [112].

For faultless impregnation (no air inclusions), the resin must be introduced into the FC component through suitable sprue channels. Dispersions in the permeability of the dry deposited fibres result in the flow front developing individually for each component. The infusion should therefore ideally be sensor-controlled (Sect. 2.6) according to the determined resin front development, i.e., the resin channels should be "switchable". At the same time, it is desired to avoid costs and waste and to use such switchable channels repeatedly for the impregnation of many components. The **vacuum differential pressure process** (Fig. 2.10) allows for selective activation of reusable distribution channels without component imprints while reducing resin consumption [48] and is described in a tutorial [28].

The autoclave is the most efficient tool for generating a high FVC for large-area components. Disadvantages are the vacuum bagging, the thermal inertia, which makes targeted control difficult, limited observability of the process material during curing, long process times (up to 9 h) and high power requirement (up to 6 MW). The latter can be largely overcome by predicting the thermal processes in the autoclave during the process time by means of a coupled thermodynamic flow simulation and by using a suitable sensor system on the component for early detection of TARGET-ACTUAL deviations and to determine the degree of cure. Up to 50% time and 30% energy savings are possible by using a **virtual autoclave** and control via dielectric sensors to determine the degree of cure [111].

To realise lightweight design with **continuous filaments in 3D printing**, the fibres must be embedded in a fusible matrix material. However, thermoplastics require special impregnation processes to evenly enclose the filaments due to their high viscosity.

With a new ultrasound-based impregnation technology (Fig. 2.11), the cost of continuous fibre-reinforced thermoplastic semi-finished products for 3D printing can be reduced by 80% [109]. For learning more on the development of continuous fibre 3D printing, cf. [34].

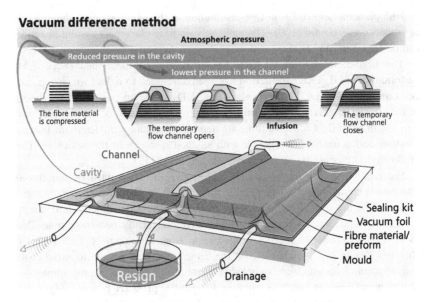

Fig. 2.10 Operating principle of the vacuum differential pressure process

2.6   New Quality Assurance Procedures

Depending on the technology and component, quality inspection of FC components can account for up to 30% of manufacturing time. Today's quality checks can be replaced by online quality control with increasing sensor quality and computing power. Online detection of deviations and quick identification of necessary corrections save considerable time and costs. Here are just a few examples of recent developments.

A typical manufacturing deviation is **fibre waviness,** which occurs when co-bonding pre-cured components with not yet cured FC shells. Understanding how these deviations interact with loads and potential damage during aircraft operation helps minimise rework or select KDFs for structural properties less conservatively. How fibre waviness interacts with impact damage under subsequent longitudinal compressive loading of an FC shell can now be described and has been validated by testing [12].

Fibre placement errors in ATL can be gaps or overlaps of the fibre tapes and, in addition, twists in the deposit with AFP. Since black fibres reflect little light,

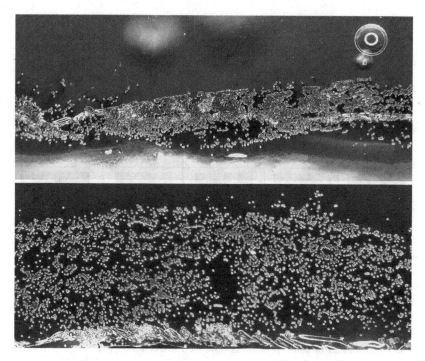

Fig. 2.11   Fibre impregnation in a melt bath a) without and b) with ultrasonic action

optical defect detection is difficult. Currently, laser line scan sensors (LLSS) are most commonly used for optical deposit monitoring (Fig. 2.12).

Today, the camera image of a laser scan can be suitably modelled, the quality of the LLSS signal evaluated for defect detection [77] and a classification with up to 100% hit rate can be realised [76].

If a deviation from manufacturing specifications is detected, an **in-situ evaluation** is required to decide directly in the process whether tolerances can be applied or corrections must be made.

The influence of a missing fibre tape during the placement of a wing shell on the stress distribution can be seen in real time by comparing a local nominal and actual FEM [45].

A possible defect in infusion processes is the retention of dry inclusions in the FC component (dry spots) due to unevenly advancing resin fronts. The resin channels attached to the component are determined by an **infusion strategy** based

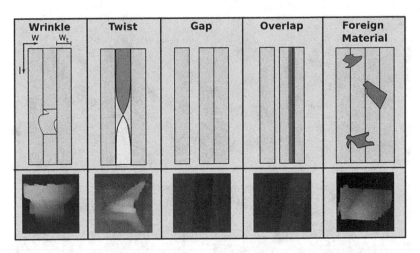

Fig. 2.12 Common fibre placement defects, schematic and exemplary as LLSS image [78]

on a precise knowledge of the flow behaviour of a resin in a cavity filled with fibre material under parameters such as pressure and temperature. How the parameters of matrix viscosity, fibre permeability and infusion pressure affect the formation of dry spots in an infusion process is now well understood [18].

For a high FVC, the FC component is completely encapsulated by vacuum bags in open moulds, a vacuum is applied, and the encapsulation is additionally pressed against the component by external pressure – 6 to 15 bar in the autoclave. For faultless compaction and – in the case of resin infusion – complete component impregnation, the vacuum bag must not contain any leaks. However, due to the size of the component and its geometric complexity, for example in the case of shell structures, leaks often occur. Fast **vacuum leak detection** is therefore a crucial cost factor. Using thermography and piezoelectric pressure sensors, leaks can be found automatically with 20% time savings compared to conventional methods [40].

Monitoring the **flow front development** of an infusion requires suitable sensors when the mould is closed. Piezoceramic ultrasonic (US) sensors are suitable for pulse-echo applications (Fig. 2.13). The density change in a fibre structure before and after impregnation with resin leads to significant amplitude jumps in the signal response. The sensors can be built very small and integrated into the mould without coming into direct contact with the resin. They can be used to

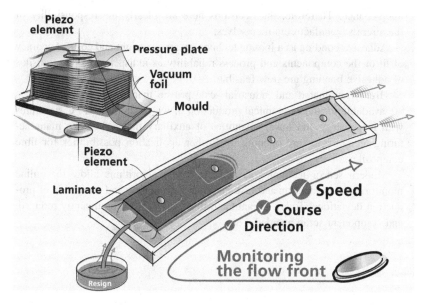

Fig. 2.13 Principle of ultrasonic sensor integration for flow front monitoring [70]

determine flow front curves very accurately, flow front velocities to within 5% and flow directions to within 19% [70].

Conclusion

New material combinations and semi-finished products are undergoing dynamic development. A major hurdle for their use in aircraft design is the extensive and thus cost-intensive qualification by specially certified testing laboratories (e.g., according to NADCAP [8], [50]). Cross-industry standards and qualification programs are needed to guarantee sufficient sales volumes of new semi-finished products.

In order to be able to build significantly lighter with FC, the detectability of damage in this class of materials must be improved. With higher allowable strains in CFRP structures, primary structure weight savings of as little as 5% to 10% could be achieved.

New designs show the greatest lightweight potential, but have far-reaching implications for manufacturing concepts, system integration and in-service maintenance. Therefore, the decisions here are clearly the responsibility of the aircraft manufacturers themselves.

Adhesive bonding as a joining technology saves weight and costs. Accuracy of fit of the components and process reliability as indispensable prerequisites of adhesive bonding are now feasible.

Highly automated and industrial-scale proven manufacturing technologies are available for the economical production of weight-optimised FC structures with less energy and lower quantities of auxiliary materials. New manufacturing technologies are opening up further application possibilities for fibre composites.

Modern sensor technologies and evaluation algorithms allow for online monitoring of production and fast, automated evaluation and correction of production deviations. Today's manufacturing costs can be significantly reduced, and component scrap avoided.

Lightweight System Design with Integration of Passive Functions

Passive functions are not used for load transfer alone, as in classic lightweight design, but fulfil other requirements for the overall product, such as minimising aerodynamic drag, providing electrical conductivity and thermal or acoustic insulation.

3.1 Structures for the Natural Laminar Flow

Laminar flow reduces frictional drag and, according to Brequet, contributes directly proportional to the reduction of an aircraft's energy consumption, Sect. 1.2.

The reduction in frictional drag due to laminar flow on the upper surface of the wing is estimated to be up to 8% [51]. Designing structures in such a way that the flow remains laminar for as long as possible therefore has a direct influence on the energy efficiency of an aircraft. Natural laminar flow (NLF) can be significantly supported by suitable shaping of the structure, in particular by avoiding gaps or discontinuities at joining edges.

Even today's CFRP wing shells are still manufactured with rivets. These cause unevenness and protrusions, which promote vortex formation and premature flow reversal. An integral **laminar upper wing shell –** taking into account PID and ensuring a high FVC – places special demands on design and manufacturing. A solution approach for a fully integral laminar CFRP wing upper shell is described using the example of a 2.5 m × 1.5 m demonstrator in [52] (Fig. 3.1).

The manufacturing-related **waviness** of a CFRP wing upper shell influences the run length of the laminar flow over the wing depth. An ideally laminar wing

© Deutsches Zentrum für Luft- und Raumfahrt e. V. (DLR), Linder Höhe, 51147 Köln 2024
M. Wiedemann, *System Lightweight Design for Aviation*, essentials,
https://doi.org/10.1007/978-3-031-44165-3_3

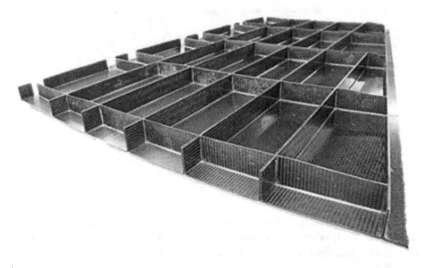

Fig. 3.1 Demonstrator panel of an integral laminar wing upper shell

should have laminar flow up to a relative wing depth of 60%. PID leads to concave waviness in the skin and load-induced deformation (LID) leads to a convex "pillow" formation. Net, a disturbing ripple remains, which increases the flow resistance of the laminar wing. The laminar run length of the flow over a wing surface can be reduced by up to 4% due to influences from PID and LID [46].

Essential for NLF is the avoidance of gaps or steps between adjacent components on the flowed- around side of a structure, cf. Fig. 3.2 left. Rivet heads of the connection of the leading edge to the wing shell also have a negative effect. In addition, there is the maintenance requirement for rapid replacement of the leading edge in the event of damage while maintaining the smallest tolerances. Therefore, a **laminar CFRP leading edge** with metallic covering foils was developed with a maximum step height < 0.15 mm [99], whose rapid replacement is ensured by a tolerance-compensating connection with eccentric bushings [107].

3.2   Electrical Conductivity of CFRP

CFRP structures have low electrical conductivity due to the matrix material. In order to realize the required electrical conductivity (Electric Structure Network: ESN), currently CFRP fuselage structures (A350 fuselage, [10]) are equipped

Fig. 3.2   Laminar wing leading edge – smoothing out unevenness through metallic cover foil

with additional non-load-bearing metal. This results in additional weights for the aircraft and additional costs in production.

At 6000 S/m, carbon fibres exhibit about 30% of the metallic conductivity, but the surrounding matrix material at about 10^{-8} S/m acts as an insulator to prevent both lightning protection and the ground connection of electrical loads. **Electrical conductivity** in CFRP laminate thickness can be increased to 600 S/m by using silver-coated polyamide filaments in combination with conductive non-woven layers [95] and reduces the required basis weight of lightning protection on a CFRP surface from 175 g/m^2 to 25 g/m^2 [94] (Fig. 3.3).

Electrical cables with a total weight of 3 tons are installed in the A380 for data communications [61]. **FML hybrids** are redundant with respect to contacting and damage; local defects cannot affect electrical conductivity. Together with mechanical advantages, low frequency electrical signals can thus be transported through the laminate on several levels. The dielectric strength of the individual layers is decisive for the permissible electrical voltage that can be applied. For metal foils electrically separated with three 0.1 mm thick glass-fibre layers and epoxy resin matrix, 250 V to 600 V are transmissible [91].

The integration of electrically conductive layers in an FC also enables targeted heating, for example, for **deicing leading edges of wings** without feeding system lines. With a suitable design implementation (Fig. 3.4), electrical deicing is possible with one third of the heat output of a conventional anti-icing system, i.e., about 3.6 kW/m^2 [89].

Cabin elements are usually made of non-conductive sandwich FC and are characterised by a large number of electrical consumers. In addition to grounding, the power supply must be installed separately with resulting additional weight and costs.

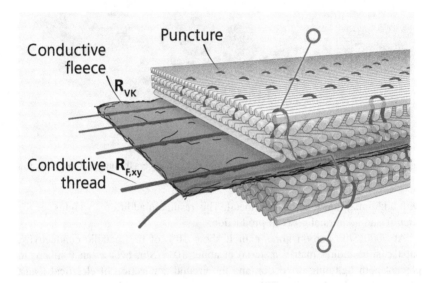

Fig. 3.3 Increasing the electrical conductivity of NCF bonding between textiles in the contact plane [94]

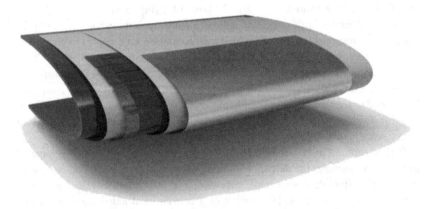

Fig. 3.4 Wing leading edge with deicing system and metallic abrasion protection

Fig. 3.5   Trace integration on the rear wall of an A330 galley

By **integrating the conductors** into the rear wall of an A330 galley, weight savings of 30% could be demonstrated compared to the current state of the art [90] (Fig. 3.5).

The integration of electrical traces into an FC structure offers a lot of potential if reliable contacting can be ensured. A multifunctional load insert for SW structures with integrated electrical signal transmission and thermal load transfer has recently been designed and successfully tested for a satellite wall panel [86].

3.3   Noise Transmission into the Cabin

CFRP fuselage structures increase sound propagation and radiation into a cabin because of the high material stiffness. Insulation material is currently used to reduce turbine and flow noise to the cabin. A lighter option is the use of a passive damping layer (Passive Constrained Layer Damping: PCLD), whose weight per unit area is lower at approx. 0.83 kg/m^2. The effect of PCLD in a rigid grid panel (comp. Sect. 2.3) on the radiated sound power was found to be -2 dB for frequencies above 300 Hz [108].

Rigid sandwich panels of an aircraft cabin are only suitable for sound insulation to a limited extent. However, the sound reduction index of these secondary structures can be increased by a suitable, mass-constant design of the honeycomb core and used specifically for acoustically adapted transmission properties.

In a simulation, it can be shown that for an SW panel with GFRP face sheets and printed plastic honeycomb core, the sound reduction index can be changed selectively [93]. Experiments show that the sound reduction index becomes effective from 500 Hz upwards, depending on the honeycomb core geometry and its cover layer support [92] (Fig. 3.6).

Conclusion

The fully integral CFRP design and modern bonding technologies allow for steps and irregularities on the aerodynamic surface of structures to be largely avoided. This enables laminar flow and effective reduction of aerodynamic drag.

Selectively increasing or decreasing the electrical conductivity of structures allows for different functions to be realised with minimum weight.

Design possibilities of the FC structure are effective up to the range of acoustic radiation.

Fig. 3.6 Different plastic honeycomb structures with different sound reduction indexes

Lightweight System Design with Integration of Active Functions

4

FC-compatible designs, structural bonding and the use of smart materials from adaptronics characterise the integration of active functions. In interaction with aerodynamics, lightweight system design can have a special effect by enabling hybrid laminar flow control. Furthermore, active reduction of sound transmission into the cabin or structural monitoring can be realised in an integrated way.

4.1 Structures for Hybrid Laminar Flow Control

The importance of laminar flow for the energy consumption of an aircraft has already been discussed in Sect. 3.1. The particular challenge is to synthesise the different requirements of a support structure with the extended requirements of, for example, active hybrid laminar flow control (HLFC). HLFC with active suction of the boundary layer flow theoretically allows for a drag reduction of 30% [17].

Already at the leading edges of the wing, turbulence starts to form under certain inflow conditions. An active suction system in the structure and a micro-perforated airfoil surface are required to actively prevent this. The supporting structure behind an aerodynamic surface must be suitably designed with chambers through which—adapted to the flow conditions around the leading edge—suction is applied with different pressure gradients, Fig. 4.1. Systematic and multidisciplinary development of an HLFC system was prototyped for the A350 horizontal tailplane (HTP) in the "Clean Sky II" ECHO project [105]. Based on a new HLFC leading edge design, a potential of 5% fuel savings was identified [39].

© Deutsches Zentrum für Luft- und Raumfahrt e. V. (DLR), Linder Höhe, 51147 Köln 2024
M. Wiedemann, *System Lightweight Design for Aviation*, essentials, https://doi.org/10.1007/978-3-031-44165-3_4

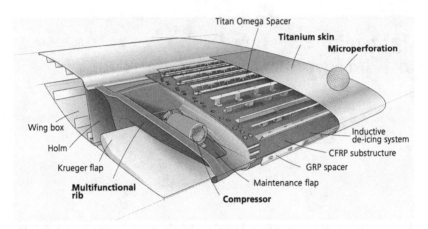

Titan Omega Spacer
Titanium skin
Microperforation
Wing box
Holm
Inductive de-icing system
CFRP substructure
Krueger flap
GRP spacer
Multifunctional rib
Maintenance flap
Compressor

Fig. 4.1   Model of an HLFC leading edge of a tailplane

4.2   Shape Variability

Contour changes of airfoils allow for adaptation to different flight conditions. However, active shape variability must count with the inherent stiffness of the structure into which it is integrated. Thus, there are limits to the resulting shape changes. Within these limits, a contour change can have an strong effect and replace otherwise necessary additional aggregates or functional elements.

Laminar airflow is not possible with conventional moving leading edges due to unavoidable steps to the wing box. In order to ensure a step- and gap-free change of the leading edge angle of attack in take-off and landing configurations, an inherently shape-variable leading edge of the wing is desirable. In the 7th EU Framework Programme, the SARISTU (Smart Intelligent Aircraft Structures) project investigated possible applications of morphing (shape variability) on leading and trailing edges. Together with Airbus and partners, a **shape-variable wing leading edge** was developed, built and experimentally investigated. The particular challenge is the simultaneous consideration of the other requirements for this component such as lightning protection, de-icing, abrasion, and bird strike protection. In an aircraft configuration with rear mounted engines, the new leading edge would save 1% fuel [59]. Newer designs, thanks to a special material hybrid, can realise a leading edge lowering of up to 20° with a 4–9% increase in lift coefficient [115], Fig. 4.2.

Fig. 4.2 "Droop Nose" demonstrator with elastomer-GFRP hybrid skin

In order to reduce the wave resistance of the transonic flow over the wing, so-called **shock control bumps** (SCBs) can be used, which cause controlled turbulence condition from a defined wing depth and with a defined height.

Thus, a shape-variable adaptive spoiler was prototypically developed, which allows to adjust the required deformability in wing depth direction and height [68], Fig. 4.3.

One way of increasing the lift of an airfoil is to exploit the Coanda effect: the flow on a flap is controlled for a longer period of time if it is fitted with an **active blow-out lip.** This requires rapidly actuable structural elements for targeted flow control, which are integrated into the flap in a small installation space.

A structurally compliant dynamic piezoelectric actuation of a blow-out lip shows a lift increase of $\Delta C_a = 0.57$ in wind tunnel measurements [119] (Fig. 4.4).

Gapless flaps or movable winglets avoid flow losses and can be used for load reduction. One way of actuating a structure with built-in joints is to apply internal pressure. **Pressure-actuated cell structures** (PACS) and hydraulically actuated compact unit structures (fluid actuated morphing unit structures: FAMoUS) allow for the realisation of large deformations, Fig. 4.5. However, the fatigue strength of

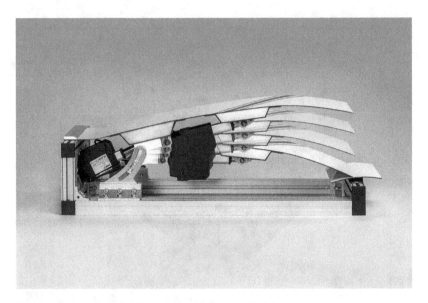

Fig. 4.3 Design of a flap with actuable shock control bump

the solid-state joints and the pressure tightness of the cells pose challenges. Here, continuous fibre-reinforced 3D printing opens up new perspectives (cf. Sect. 2.5).

For a wing airfoil, a trailing edge lowering of 15° and thus a theoretical increase in lift by a factor of 3 was demonstrated by a PACS design [36]. A shape-variable winglet with structurally conformal actuation was developed and tested in the wind tunnel as part of the EU project NOVEMORE [114].

4.3   Vibration Influence

Vibrations emanating from propellers and turbines cause material fatigue, wear and comfort restrictions. **Active vibration reduction** is a major application area of lightweight system design, in that structurally integrated actuators counteract the resulting deformations at the same frequency. Two CFRP rods with an integrated piezo stack actuator in a truss structure, similar to an engine suspension, controlled by an adaptive controller, reduce amplitudes by 40 dB [101]. In a truss-work, Fig. 4.6, the vibration transmission of a propeller to the support structure is reduced by 80% to 90% [102].

Fig. 4.4 Active blow-out lips for utilising the Coanda effect in the wind tunnel

Fig. 4.5 Pressure-actuated cell structures for variable shape wing structures

Ice buildup on the leading edges of aerodynamic surfaces increases flow resistance. Beyond a certain ice adhesion, this becomes critical to safety and must be detected and dissolved. By integrating suitable active elements into the supporting structure of the leading edge of the wing, both safe and fast **ice detection** and mechanical deicing—instead of thermal deicing, cf. Sect. 3.2—can be realised.

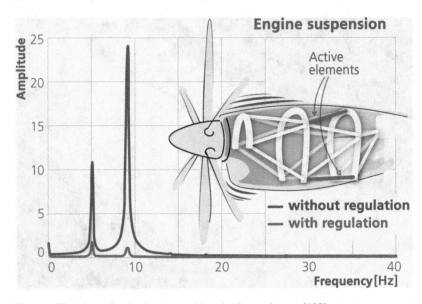

Fig. 4.6 Vibration reduction in a truss with active beam element [103]

Integrated piezo actuators allow for detection of local ice adhesions on leading edges of wings from 2 mm thickness by means of ultrasonic signals [79]. Investigations with integrated electro-mechanical systems show a potential for **active deicing** by means of local, high-frequency skin deformation [33], Fig. 4.7.

Structural vibrations from engines and flow turbulence on the fuselage are transmitted to the cabin, where they are perceived as noise. In addition to heavy insulating material, counter-sound techniques (Active Noise Control: ANC) now minimize sound radiation.

A lighter and more efficient system takes advantage of the fact that sound waves can no longer be radiated by a structure if the structure vibrates below the so-called coincidence frequency in wavelengths that are smaller than the wavelengths of the radiated sound waves (Active Structure Acoustic Control: ASAC), the so-called acoustic short circuit. More information on the ASAC method in [82].

With the aid of an ASAC system integrated into the cabin lining, multitone, low-frequency interference excitations can be reduced by up to 20 dB [81]. A ready-to-install **active cabin panel** that integrates an ASAC system reduces the sound pressure level for a turboprop cabin by 6.8 dB [80], Fig. 4.8.

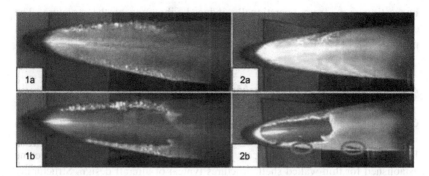

Fig. 4.7 Structure-integrated deicing in ice wind tunnel; ice buildup (a) before and (b) after deicing after 4 min at −10 °C (1) and after 2 min at −20 °C (2); [33]

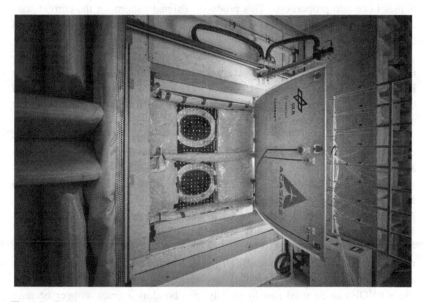

Fig. 4.8 Cabin panel with ASAC system in transmission test rig

The wireless use of sensors for flight condition monitoring saves manufacturing costs and weight if energy is locally available. For a wireless sensor network (WSN), the necessary operating energy can also be generated by **energy** harvesting, e.g., from operationally induced vibrations of the structure. The energy demand of autonomous sensor elements can be met by structure-integrated piezoceramics [37].

4.4 Structural Health Monitoring—SHM

Continuous monitoring of the structural integrity of lightweight structures has a major influence on structural properties and resulting design weight when using FC components, Sect. 2.2. In recent years, the **Lamb wave method** has become established for thin-walled shell structures typical of aircraft design. Structurally integrated (or applied) piezoelectric elements are used to establish an actuator-sensor network that is used to transmit high-frequency longitudinal and transverse waves through the shells. These are reflected and transmitted at stiffness disconti-nuities in certain proportions. This produces a signal pattern for the current state of a shell, which, compared to a stored pattern for the intact state, can be used to infer locations of damage and damage magnitudes [85].

On a CFRP door frame shell (Fig. 4.9), a network of 584 piezoelectric ele-ments is able to localise visually barely detectable delaminations (Barely Visible Impact Damages: BVID) from 310 mm^2 to 2311 mm^2 with an accuracy of 5 mm to 85 mm [83]. Lamb-wave SHM can also be applied for damage detection over a temperature range of -42 °C to 85 °C [84].

For the evaluation of a detected structural damage with respect to load-carrying capacity and a possible need for repair, an automated comparison with a simulation is required, which analyses the residual bearing behaviour con-sidering the detected damage. Linking of Lamb-wave-based damage detection and **damage assessment** becomes possible by using fast surrogate models [35]. Lamb-wave SHM can also be used in aircraft maintenance in combination with simulation and **augmented reality** [120].

One possibility for damage detection in joints is offered by film sensors made of polyvinylidene fluoride (PVDF), which also have piezoelectric properties and allow for strains to be measured very accurately with suitable pretreatment. Since PVDF can be used simultaneously in a bond as a crack stopper because of its toughness, Fig. 4.10, such sensors also lend themselves to monitoring bonded joints. A 100 μm thick PVDF film as a crack stopper with applied metal measuring grid of 200 nm allows sensing of strains in an adhesive seam [41].

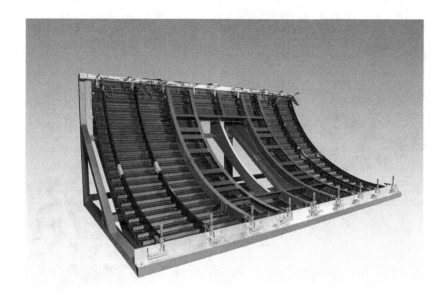

Fig. 4.9 CFRP door frame shell with SHM network for automated damage detection in the EU "Clean Sky II" project SARISTU

4.5 Structural Batteries

In the context of increasing electrification of on-board functions and drives, the need for batteries for intermediate storage is growing. Battery storage can also be designed as load-bearing structures and used for load transfer. The more mechanical load it carries as part of the structure, the more its weight share as a separate battery decreases. Decisive for the structural load-bearing capacity is the use of solid-state electrolytes as storage medium and suitable electrical contacting with maximum surface area for the best possible capacitive energy storage. While solid-state electrolytes (e.g., $Li_{1+x}Al_xTi_{2-x}(PO_4)_3$: LATP) are required, carbon nanotubes (CNT) are suitable for electrical contacting. Thus, parts of the secondary structure of an aircraft could be used to store a few kWh of electrical energy, Fig. 4.11.

By combining an LATP with CNTs and a suitable structural design, a storage capability of 11.59 mF/cm^3 can be demonstrated [69].

A structure-integrated capacitive element in an aerospace application for energy storage from the deceleration of rotating masses could be built 73% lighter

Fig. 4.10   Sensor inlay for crack detection in glued seams

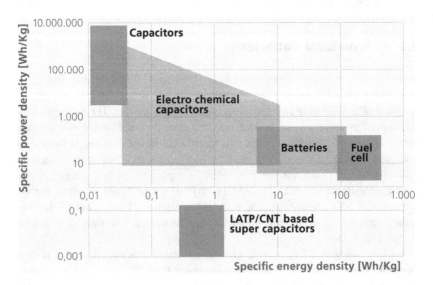

Fig. 4.11   Comparison of power and energy density for different energy storage devices

and 78% smaller than the classic comparative structure with pure batteries. The mechanical load capacity was retained by 80% in the process [87].

Conclusion

The integration of active functions into the load-bearing structure offers further and diverse potential for weight and drag reduction. However, developments in this field require a paradigm shift in aircraft certification, as systems that were previously approved separately must be considered integrally. In this field of lightweight system design, a new quality of interdisciplinary cooperation is required for success.

What you can take away from this *essential*

- A definition of lightweight system design
- A presentation of potentials for emission-minimising aviation
- A small collection of technology examples
- Suggestions for further developments in the field of lightweight system design

© Deutsches Zentrum für Luft- und Raumfahrt e.V. (DLR), Linder Höhe, 51147 Köln 2024
M. Wiedemann, *System Lightweight Design for Aviation*, essentials,
https://doi.org/10.1007/978-3-031-44165-3

Literature

1. (2016) Integration von Power to Gas/Power to Liquid in den laufenden Transformationsprozess. https://www.umweltbundesamt.de/sites/default/files/medien/1/publik ationen/position_power_to_gas-power_to_liquid_web.pdf. Accessed 31 Mar. 2021
2. (2017) Strom 2030. Langfristige Trends—Aufgaben für die kommenden Jahre. Ergebnispapier. https://www.bmwi.de/Redaktion/DE/Publikationen/Energie/strom-2030-erg ebnispapier.pdf?__blob=publicationFile&v=32. Accessed 8 Apr. 2021
3. (2020) Breguet'sche Reichweitenformel. https://de.wikipedia.org/w/index.php?title= Breguet'sche_Reichweitenformel&oldid=207061489. Accessed 25 July 2021
4. (2020) Kosten und Transformationspfade für strombasierte Energieträger. Studie im Auftrag des Bundesministeriums für Wirtschaft und Energie. Accessed 30 Mar. 2021
5. (2020) Wasserstoff-und-wasserstoffbasierte-Brennstoffe. Eine Überblicksuntersuchung. https://www.oeko.de/fileadmin/oekodoc/Wasserstoff-und-wasserstoffbasi erte-Brennstoffe.pdf. Accessed 8 Apr. 2021
6. (2021) DLR—Institut für Faserverbundleichtbau und Adaptronik—Innovationsberichte. https://www.dlr.de/fa/desktopdefault.aspx/tabid-10605/. Accessed 25 Sept. 2021
7. (2021) DLR—Zentrum für Leichtbauproduktionstechnologie—Standort Stade. https:// www.dlr.de/zlp/desktopdefault.aspx/tabid-10811/#gallery/25907. Accessed 29. Juni 2021
8. (2021) Nadcap. https://en.wikipedia.org/w/index.php?title=Nadcap&oldid=104018 9287. Accessed 25 Sept. 2021
9. Air Liquide Energies (2017) How is hydrogen stored? Air Liquide Energies
10. Airbus (2014) Flight airworthiness authority technology—FAST. No. 54. Airbus Technival Magazine, January
11. Airbus (2021) Extended Service Goal (ESG). https://services.airbus.com/en/flight-operations/system-upgrades/operations-extension/extended-service-goal-esg.html. Accessed 8 July 2021
12. Al-kathemi N, Wille T, Heinecke F, Degenhardt R, Wiedemann M (2021) Interaction effect of out of plane waviness and impact damages on composite structures—an experimental study. Compos Struct 276:114405. https://doi.org/10.1016/j.compstruct.2021. 114405

© Deutsches Zentrum für Luft- und Raumfahrt e.V. (DLR), Linder Höhe, 51147 Köln 2024
M. Wiedemann, *System Lightweight Design for Aviation*, essentials, https://doi.org/10.1007/978-3-031-44165-3

13. Amacher R, Cugnoni J, Botsis J, Sorensen L, Smith W, Dransfeld C (2014) Thin ply composites: experimental characterization and modeling of size-effects. Compos Sci Technol 101:121–132. https://doi.org/10.1016/j.compscitech.2014.06.027

14. Bachmann J, YI X (Hrsg) (2019) EU/China research on ECO-COMPOSITES for Aviation Interior and Secondary Structures. https://elib.dlr.de/130971/

15. Bachmann J, Wiedemann M, Wierach P (2018) Flexural mechanical properties of hybrid epoxy composites reinforced with nonwoven made of flax fibres and recycled carbon fibres. Aerospace 5(4):107. https://doi.org/10.3390/aerospace5040107

16. Bauhaus Luftfahrt (2021) Die Grenzen der Batterietechnologie. https://www.bauhaus-luftfahrt.net/de/forschung/energietechnologien-antriebssysteme/die-grenzen-der-batterietechnologie/. Accessed 8 Apr. 2021

17. Beck N, Landa T, Seitz A, Boermans L, Liu Y, Radespiel R (2018) Drag reduction by laminar flow control. Energies 11(1):252. https://doi.org/10.3390/en11010252

18. Bertling D, Kaps R, Mulugeta E (2016) Analysis of dry-spot behavior in the pressure field of a liquid composite molding process. CEAS Aeronaut J 7(4):577–585. https://doi.org/10.1007/s13272-016-0207-2

19. Bogenfeld R, Kreikemeier J, Wille T (2018) Review and benchmark study on the analysis of low-velocity impact on composite laminates. Eng Fail Anal 86:72–99. https://doi.org/10.1016/j.engfailanal.2017.12.019

20. Bogenfeld RM (2019) A combined analytical and numerical analysis method for low-velocity impact on composite structures. Dissertation, Technischen Universität Carolo-Wilhelmina

21. Boose Y, Kappel E, Stefaniak D, Prussak R, Pototzky A, Weiß L (2020) Phenomenological investigation on crash characteristics of thin layered CFRP-steel laminates. International Journal of Crashworthiness:1–10. https://doi.org/10.1080/13588265.2020.1787681

22. Buelow C, Heltsch N, Hirano Y, Aoki Y, Kawabe K (2017) Investigation of the Impact Properties of Thin-Ply Prepreg at Element Level. Proceedings of SAMPE Japan 2017. https://elib.dlr.de/122343/

23. Bundesministerium für Verkehr und digitale Infrastruktur (2020) Werkstattbericht alternative kraftstoffe. Klimawirkungen und wege zum einsatz alternativer kraftstoffe

24. Camanho PP, Fink A, Obst A, Pimenta S (2009) Hybrid titanium–CFRP laminates for high-performance bolted joints. Compos A Appl Sci Manuf 40(12):1826–1837. https://doi.org/10.1016/j.compositesa.2009.02.010

25. Das S (2011) Life cycle assessment of carbon fiber-reinforced polymer composites. Int J Life Cycle Assess 16(3):268–282. https://doi.org/10.1007/s11367-011-0264-z

26. Delisle DPP, Schreiber M, Krombholz C, Stüve J (2018) Fertigung von Faserverbundstrukturen mittels kooperierender Robotereinheiten. Lightweight Design (2/2018)

27. Dienel CP, Meyer H, Werwer M, Willberg C (2019) Estimation of airframe weight reduction by integration of piezoelectric and guided wave–based structural health monitoring. Struct Health Monit 18(5–6):1778–1788. https://doi.org/10.1177/1475921718813279

28. DLR Leichtbau (2021) Leichtbau Tutorial #02 Was die Vakuuminfusion mit dem Trinken eines Cocktails zu tun hat. https://leichtbau.dlr.de/leichtbau-turtorial-02. Accessed 25 Sept. 2021

29. Düring D, Weiß L, Stefaniak D, Jordan N, Hühne C (2015) Low-velocity impact response of composite laminates with steel and elastomer protective layer. Compos Struct 134:18–26. https://doi.org/10.1016/j.compstruct.2015.08.001

30. Düring D, Petersen E, Stefaniak D, Hühne C (2020) Damage resistance and low-velocity impact behaviour of hybrid composite laminates with multiple thin steel and elastomer layers. Compos Struct 238:111851. https://doi.org/10.1016/j.compst ruct.2019.111851

31. EASA (2021) EASA Search I EASA. https://www.easa.europa.eu/search?keys=AMC+ 20-29+. Accessed 27 June 2021

32. Eidgenössisches Departement für Umwelt, Verkehr (2020) Faktenmaterial Elektrisches Fliegen. https://www.bazl.admin.ch/dam/bazl/de/dokumente/Politik/Umwelt/ faktenblatt_elektrisches_fliegen.pdf.download.pdf/Faktenmaterial%20Elektrisches% 20Fliegen.pdf. Accessed 8 Apr. 2021

33. Endres M, Sommerwerk H, Mendig C, Sinapius M, Horst P (2017) Experimental study of two electro-mechanical de-icing systems applied on a wing section tested in an icing wind tunnel. CEAS Aeronaut J 8(3):429–439. https://doi.org/10.1007/s13272-017-0249-0

34. FA-Podcast (2021) Episode 2. https://leichtbau.dlr.de/episode-2. Accessed 25 Sept. 2021

35. Garbade M (2018) Efficient simulation of the through-the-thickness damage composition in composite aircraft structures for use with integrated SHM systems. 9th International Conference on Computational Methods (ICCM2018). https://elib.dlr.de/124 721/

36. Gramüller B (2016) On pressure-actuated cellular structures. Dissertation, Technische Universität Carolo-Wilhelmina Braunschweig

37. Grasböck L, Humer A, Nader M, Schagerl M, Mayer D, Misol M, Humer C, Herold S, Monner HP (Hrsg) (2019) Wireless sensor networks and energy harvesting for energy autonomous smart structures. 4SMARTS 2019. Shaker

38. Graver B (2021) CO2 emissions from commercial aviation: 2013, 2018, and 2019 I International Council on Clean Transportation. https://theicct.org/publications/co2-emi ssions-commercial-aviation-2020. Accessed 29 Nov. 2021

39. Haase T, Ropte S, van Kamp de B, Pohya AA, Kleineberg M, Schröder A, Pauly J-L, Kilian T, Wild J, Herrmann U (2020) Next generation wings for long range aircraft: hybrid laminar flow control technology drivers. Deutscher Luft- und Raumfahrt Kongress

40. Haschenburger A, Menke N, Stüve J (2021) Sensor-based leakage detection in vacuum bagging. Int J Adv Manuf Technol 116(7–8):2413–2424. https://doi.org/10.1007/s00 170-021-07505-5

41. Heide C von der, Steinmetz J, Schollerer MJ, Hühne C, Sinapius M, Dietzel A (2021) Smart inlays for simultaneous crack sensing and arrest in multifunctional bondlines of composites. Sensors (Basel) 21(11). https://doi.org/10.3390/s21113852

42. Heilmann L (2020) Fused Bonding—Zuverlässiges Kleben durch reaktionsfähige Oberflächen. Innovationsbericht. https://elib.dlr.de/136021/

43. Heilmann L (2020) Qualitätskontrolle von Reparaturklebungen an Faserverbundstrukturen durch vollflächige Festigkeitsprüfung. Dissertation, Technische Universität Carolo-Wilhelmina Braunschweig

44. Heilmann L (2021) Leichtbau Tutorial #1 Das Kleben von Faserverbundstoffen und das Fused Bonding-Verfahren. https://leichtbau.dlr.de/leichtbau-tutorial-1-das-kleben-von-faserverbundstoffen-und-das-fused-bonding-verfahren. Accessed 27. June 2021
45. Heinecke F, Wille T (2018) In-situ structural evaluation during the fibre deposition process of composite manufacturing. CEAS Aeronaut J 9(1):123–133. https://doi.org/10.1007/s13272-018-0284-5
46. Heinrich L, Kruse M (2016) Laminar composite wing surface waviness—two counteracting effects and a combined assessment by two methods. Deutscher Luft- und Raumfahrtkongress
47. Herrmann R (2021) Final report summary—MAAXIMUS (More Affordable Aircraft structure through eXtended, Integrated, and Mature nUmerical Sizing). Publication Office/CORDIS. https://cordis.europa.eu/project/id/213371/reporting. Accessed 6 Sept. 2021
48. Hindersmann A (2017) Beitrag zur Simulation und Verbesserung der Vakuumdifferenzdruckinfusion. Deutsches Zentrum für Luft- und Raumfahrt e. V
49. Hindersmann A (2019) Confusion about infusion: an overview of infusion processes. Compos A Appl Sci Manuf 126:105583. https://doi.org/10.1016/j.compositesa.2019.105583
50. Hoidis J (2021) DLR—Institut für Faserverbundleichtbau und Adaptronik—NADCAP-akkreditierte Materialprüfung am Institut. https://www.dlr.de/fa/desktopdefault.aspx/tabid-10731/13411_read-58144/. Accessed 24 Sept. 2021
51. Horst P, Elham A, Radespiel R (2021) Reduction of aircraft drag, loads and mass for energy transition in aeronautics. https://doi.org/10.25967/530164
52. Hühne C, Ückert C, Steffen O (2015) Entwicklung einer laminaren Flügelschale. Lightweight Des 8(6):32–37. https://doi.org/10.1007/s35725-015-0054-9
53. Johanning A, Scholz D (2014) Conceptual aircraft design based on life cycle assessment. St. Petersburg, Russia
54. Jux M, Finke B, Mahrholz T, Sinapius M, Kwade A, Schilde C (2017) Effects of Al(OH)O nanoparticle agglomerate size in epoxy resin on tension, bending, and fracture properties. J Nanopart Res 19(4). https://doi.org/10.1007/s11051-017-3831-9
55. Kappel E (2013) Process distortions in composite manufacturing—from an experimental characterization to a prediction approach for the global scale. Dissertation, Otto-von-Guerike Universität
56. Kappel E (2018) Compensating process-induced distortions of composite structures: a short communication. Compos Struct 192:67–71. https://doi.org/10.1016/j.compstruct.2018.02.059
57. Kappel E (2022) Double–Double laminates for aerospace applications—finding best laminates for given load sets. Composites Part C: Open Access 8 (2022) 100244. https://doi.org/10.1016/j.jcomc.2022.100244
58. Kappel E, Stefaniak D, Fernlund G (2015) Predicting process-induced distortions in composite manufacturing—a pheno-numerical simulation strategy. Compos Struct 120(120):98–106. https://doi.org/10.1016/j.compstruct.2014.09.069
59. Kintscher M, Kirn J, Storm S, Peter F, Wölcken PC, Papadopoulos M (2015) Smart Intelligent Aircraft Structures (SARISTU). Proceedings of the Final Project Conference. Smart Intelligent Aircraft Structures (SARISTU):113–140. https://doi.org/10.1007/978-3-319-22413-8

60. Kleineberg M, Kaps R, Kappel E, Liebers N, Opitz S, Hein R, Azeem S (2020) EFFEKT Schlussbericht—Förderprogramm: luftfahrtforschungsprogramm LuFo V-2. Laufzeit: 01.11.2015–30.09.2019

61. Klimaschutz-Portal (2020) Das Gewicht von Flugzeugen wird immer geringer— Klimaschutz-Portal. https://www.klimaschutz-portal.aero/verbrauch-senken/am-flu gzeug/gewicht-einsparen/. Accessed 8. July 2021

62. Knote A, Ströhlein T (2009) Development of an innovative composite door surround structure for a future airliner, IB 131-2009/24. https://elib.dlr.de/59144/

63. Kolesnikov B, Herbeck L (2004) Carbon fiber composite airplane fuselage: concept and analysis. In: Central Aerohydrodinamical Institute TsAGI Russia (Hrsg) Merging the efforts: Russia in European Research Programs on Aeronautics (CD), S 1–11. https://elib.dlr.de/49781/

64. Kost C, Schlegl T (2018) Stromgestehungskosten erneuerbare Energien. https://www. ise.fraunhofer.de/content/dam/ise/de/documents/publications/studies/DE2018_ISE_ Studie_Stromgestehungskosten_Erneuerbare_Energien.pdf. Accessed 8 Apr. 2021

65. Krause D, Wille T, Miene A, Büttemeyer H, Fette M (2019) Numerical material property characterization of long-fiber-SMC materials. International Workshop on Aircraft System Technologies

66. Krombholz C, Delisle D, Perner M (2013) Advanced automated fibre placement. Advances in Manufacturing Technology XXVII. Cranfield University, UK, Cranfield University Press, S 411–416

67. Kühn A (2014) Verbesserung des Brandverhaltens von Injektionsharzsystemen durch Verwendung von Nanopartikeln. Dissertation, Technische Universität Carolo-Wilhelmina Braunschweig

68. Künnecke SC, Kintscher M, Riemenschneider J (2020) Structural design of a shock control bump for a natural laminar flow aircraft Wing. ASME Conference 2020

69. Liao G, Mahrholz T, Geier S, Wierach P, Wiedemann M (2018) Nanostructured all-solid-state supercapacitors based on NASICON-type Li1.4Al0.4Ti1.6(PO4)3 electrolyte. J Solid State Electrochem 22(4):1055–1061. https://doi.org/10.1007/s10008-017-3849-z

70. Liebers N (2018) Ultraschallsensorgeführte Infusions- und Aushärteprozesse für Faserverbundkunststoffe. Dissertation, Technische Universität Carolo-Wilhelmina Braunschweig

71. Liebisch M, Hein R, Wille T (2018) Probabilistic process simulation to predict process induced distortions of a composite frame. CEAS Aeronaut J 9(4):545–556. https://doi. org/10.1007/s13272-018-0302-7

72. Liebisch M, Wille T, Balokas G, Kriegesmann B (2019) Robustness analysis of CFRP structures under thermomechanical loading uncluding manufacturing defects. 9th EASN International Conference on Innovation in Aviation & Space. https://elib. dlr.de/130913/

73. Löbel T, Holzhüter D, Sinapius M, Hühne C (2016) A hybrid bondline concept for bonded composite joints. Int J Adhes Adhes 68:229–238. https://doi.org/10.1016/j.ija dhadh.2016.03.025

74. Lorsch P (2016) Methodik für eine hochfrequente Ermüdungsprüfung an Faserverbundwerkstoffen. Dissertation, Technische Universität Carolo-Wilhelmina Braunschweig

75. Mahrholz T, Exner W, Lorsch P, Adam J (2019) Verbundvorhaben LENAH: lebens-dauererhöhung und Leichtbauoptimierung durch nanomodifizierte und hybride Werk-stoffsysteme im Rotorblatt—Teilprojekt Materialforschung. Erfolgskontrollbericht. DLR-IB-FA-BS-2019-134

76. Meister S, Wermes M, Stüve J, Groves RM (2021) Cross-evaluation of a parallel operating SVM—CNN classifier for reliable internal decision-making processes in composite inspection. J Manuf Syst 60:620–639. https://doi.org/10.1016/j.jmsy.2021.07.022

77. Meister S, Grundhöfer L, Stüve J, Groves RM (2021) Imaging sensor data modelling and evaluation based on optical composite characteristics. Int J Adv Manuf Technol. https://doi.org/10.1007/s00170-021-07591-5

78. Meister S, Wermes MAM, Stüve J, Groves RM (2021) Review of image segmenta-tion techniques for layup defect detection in the Automated Fiber Placement process. J Intell Manuf. https://doi.org/10.1007/s10845-021-01774-3

79. Mendig C, Riemenschneider J, Monner HP, Vier LJ, Endres M, Sommerwerk H (2018) Ice detection by ultrasonic guided waves. CEAS Aeronaut J 9(3):405–415. https://doi.org/10.1007/s13272-018-0289-0

80. Misol M (2020) Active sidewall panels with virtual microphones for aircraft interior noise reduction. Appl Sci 10(19):6828. https://doi.org/10.3390/app10196828

81. Misol M, Haase T, Algermissen S, Papantoni V, Monner HP (2017) Lärmreduktion in Flugzeugen mit aktiven Linings. In: Wiedemann M, Melz T (Hrsg) Smarte Strukturen und Systeme. Tagungsband des 4SMARTS-Symposiums, 21.–22. Juni 2017, Braun-schweig. Shaker Verlag, Aachen, S 329–339

82. Misol M, Algermissen S, Haase T (2019) active control of sound, applications of encyclopedia of continuum mechanics. Springer, Berlin, S 1–13

83. Moix-Bonet M, Schmidt D, Wierach P (2018) structural health monitoring on the SARISTU full scale door surround structure. Lamb-wave based structural health mon-itoring in polymer composites. https://doi.org/10.1007/978-3-319-49715-0

84. Moix-Bonet M, Eckstein B, Wierach P (2018) Temperature compensation for dam-age detection in composite structures using guided waves. 9th European Workshop on Structural Health Monitoring, https://elib.dlr.de/123729/

85. Moll J, Kexel C, Kathol J, Fritzen C-P, Moix-Bonet M, Willberg C, Rennoch M, Koerdt M, Herrmann A (2020) Guided waves for damage detection in complex composite structures: the influence of omega stringer and different reference damage size. Appl Sci 10(9):3068. https://doi.org/10.3390/app10093068

86. Montano Rejas Z, Keimer R, Geier S, Lange M, Mierheim O, Petersen J, Pototzky A, Wolff J (2021) Design and manufacturing of a multifunctional, highly integrated satel-lite panel structure 16th European Conference on Spacecraft Structures, Materials and Environmental Testing (ECSSMET 2021)

87. Petersen J, Geier S, Wierach P (2021) Integrated thin film supercapacitor as multifunc-tional sensor system. ASME Conference 2021

88. Pfannkuche H, Bülow C (2020) Untersuchung des Verhaltens und der Verarbeitung von Thin-Ply Prepreg mittels Fertigungsversuchen und Parameterstudien. https://elib.dlr.de/138562/

89. Pototzky A, Düring D, Hühne C (2015) Entwicklung einer elektrisch betriebenen Flügelvorderkantenheizung in einem Laminarflügel. Deutscher Luft- und Raumfahrt Kongress 2015.

90. Pototzky A, Wolff J, Holzhüter D, Hühne C (2016) Abschlussbericht zum Teilprojekt des DLR im Verbund InGa (Innovative Galley). https://elib.dlr.de/104200/

91. Pototzky A, Stefaniak D, Hühne C (2018) Potentials of load carrying, structural integrated conductor tracks SAMPE Europe Conference & Exhibition 2017 Stuttgart. Stuttgart, Germany, 14–16 November 2017. Curran Associates Inc, Red Hook

92. Radestock M, Haase T, Monner HP (2019) Experimental transmission loss investigation of sandwich panels with different honeycomb core geometries 48th International Congress and Exhibition on Noise Control Engineering, INTER-NOISE 2019

93. Radestock M, Haase T (2019) Lärmtransmission durch Sandwichplatten mit verschiedenen Wabenkerngeometrien. https://elib.dlr.de/127442/

94. Rehbein J (2017) Erhöhte Blitzschlagresistenz von Kohlenstofffaserverbunden durch leitfähige Nähfäden. Dissertation, Technische Universität Carolo-Wilhelmina Braunschweig

95. Rehbein J, Wierach P, Gries T, Wiedemann M (2017) Improved electrical conductivity of NCF-reinforced CFRP for higher damage resistance to lightning strike. Compos A Appl Sci Manuf 100:352–360. https://doi.org/10.1016/j.compositesa.2017.05.014

96. Reinhard B, Torstrick S, Stüve J (2017) Automated net-shape preforming of CFRP Frames in the project Maaximus. https://elib.dlr.de/112660/

97. Rossow C-C, Geyr H von, Hepperle M (2016) The 1g-Wing, visionary concept or naive solution? Interner Bericht. https://elib.dlr.de/105029/

98. Saito H, Takeuchi H, Kimpara I A study of crack suppression mechanism of thin-ply carbon-fiber-reinforced polymer laminate with mesoscopic numerical simulation. J Compos Mater 48:2085–2096

99. Schollerer MJ, Ueckert C, Huehne C (2021) Laminar interface concept for a HLFC Horizontal Tailplane leading edge, design and manufacturing approach. Zenodo. https://doi.org/10.5281/ZENODO.4655818

100. Schreiber M, Delisle DPP (2018) Efficient CFRP-manufacturing using multiple industrial robots. https://elib.dlr.de/124521/

101. Schuetze R, Goetting HC (1996) Adaptive lightweight CFRP strut for active vibration damping in truss structures. J Intell Mater Syst Struct 7

102. Schuetze R, Goetting HC, Breitbach E, Grützmacher T (1998) Lightweight engine mounting based on adaptive CFRP struts for active vibration suppression. Aerosp Sci Technol 2(6):381–390. https://doi.org/10.1016/S1270-9638(99)80026-1

103. Schütze R, Götting C, Breitbach E, Grützmacher T (1998) Leichte CFK-Triebwerksaufhängung mit aktiver Schwingungsunterdrückung

104. Silberhorn D, Atanasov G, Walther J-N, Zill T (2019) Assessment of hydrogen fuel tank integration at aircraft level. Deutscher Luft- und Raumfahrtkongress

105. Srinivasan K, Bertram O (2019) Preliminary design and system considerations for an active hybrid laminar flow control system. Aerospace 6(10):109. https://doi.org/10.3390/aerospace6100109

106. Stefaniak D, Prussak R, Weiß L (2017) Spezifische Herausforderungen für den Einsatz von Faser-Metall-Laminaten. Lightweight Des 10(5):24–31. https://doi.org/10.1007/s35725-017-0046-z

107. Steffen O, Ückert C, Kappel E, Bach T, Hühne C (2016) A multi-material, multi-functional leading edge for the laminar flow wing. 27th SICOMP Conference

108. Titze M, Misol M, Monner HP (2019) Examination of the vibroacoustic behavior of a grid-stiffened panel with applied passive constrained layer damping. J Sound Vib 453:174–187. https://doi.org/10.1016/j.jsv.2019.03.021

109. Titze M, Opitz S, Grohmann Y, Rege M (2020) 3D-gedruckte CFK-Bauteile—Eine neue Imprägniertechnologie senkt die Kosten. Innovationsbericht. https://elib.dlr.de/136045/

110. Torstrick-v.d.Lieth S, Hessen I (2017) Kleine Serien ganz groß—Wie kann Vollautomation bei kleinen Stück-zahlen wirtschaftlich funktionieren? Innovationsbericht. https://elib.dlr.de/117372/

111. Ucan H, Scheller J, Nguyen C, Nieberl D, Beumler T, Haschenburger A, Meister S, Kappel E, Prussak R, Deden D, Mayer M, Zapp P, Pantelelis N, Hauschild B, Menke N (2019) Automated, quality assured and high volume oriented production of fiber metal laminates (FML) for the next generation of passenger aircraft fuselage shells. Sci Eng Compos Mater 26:502–508

112. Ucan H, Apmann H, Graßl G, Krombholz C, Fortkamp K, Nieberl D, Schmick F, Nguyen C, Akin D (2019) Production technologies for lightweight structures made from fibre–metal laminates in aircraft fuselages. CEAS Aeronaut J 10(2):479–489. https://doi.org/10.1007/s13272-018-0330-3

113. Vasco de Oliveira Fernandes Lopes , João (2010) Life cycle assessment of the airbus A330-200 Aircraft. Dissertation, Universidade Técnica de Lisboa

114. Vasista S, Riemenschneider J, Mendrock T, Monner HP Pressure-driven morphing devices for 3D shape changes with multiple degrees-of-freedom. ASME Conference 2018

115. Vasista S, Riemenschneider J, Keimer R, Monner HP, Nolte F, Horst P (2019) Morphing wing droop nose with large deformation: ground tests and lessons learned. Aerospace 6(10):111. https://doi.org/10.3390/aerospace6100111

116. Voß A, Handojo V, Weiser C, Niemann S (2020) Preparation of loads and aeroelastic analyses of a high altitude, long endurance, solar electric aircraft. AEC2020 Aerospace Europe Conference

117. Wiedemann M, Reinhard B (2016) RTM 4.0—Quo Vadis? Aachen-Dresden-Denkendorf International Textile Conference. https://elib.dlr.de/107957/

118. Wiedemann M, Ströhlein T, Kolesnikov B, Hühne C (2011) CFK Rumpfbauweisen. https://elib.dlr.de/71814/

119. Wierach P, Petersen J, Sinapius M (2020) Design and experimental characterization of an actuation system for flow control of an internally Blown Coanda Flap. Aerospace 7(3):29. https://doi.org/10.3390/aerospace7030029

120. Willberg C, Meyer H, Freund S, Moix-Bonet M, Dienel CP, Baalbergen E, Grooteman F, Kier T, Schulz S (2021) Process and methods for E2E maintenance architecture. Clean Sky 2 Technology Progress Review. https://elib.dlr.de/141542/

121. Wille T (2019) SuCoHS Project—Sustainable cost efficient high performance composite structures demanding temperature or fire resistance. https://elib.dlr.de/131294/

122. YouTube (2021) Projekt PROTEC-NSR. https://www.youtube.com/watch?v=6MKn_5kuuwY. Accessed 25 Sept. 2021

Printed in the United States
by Baker & Taylor Publisher Services